Hydrostatic Testing, Corrosion, and Microbiologically Influenced Corrosion

A Field Manual for Control and Prevention

Hydrostatic Testing, Corrosion, and Microbiologically Influenced Corrosion
A Field Manual for Control and Prevention

Reza Javaherdashti and Farzaneh Akvan

CRC Press
Taylor & Francis Group
Boca Raton London New York

CRC Press is an imprint of the
Taylor & Francis Group, an **informa** business

CRC Press
Taylor & Francis Group,
6000 Broken Sound Parkway NW, Suite 300,
Boca Raton, FL 33487-2742

© 2017 by Taylor & Francis Group, LLC
CRC Press is an imprint of Taylor & Francis Group, an Informa business

No claim to original U.S. Government works

Printed on acid-free paper

International Standard Book Number-13: 978-1-138-03513-3 (Paperback)
978-1-138-06006-7 (Hardback)

Visit the Taylor & Francis Web site at
http://www.taylorandfrancis.com

and the CRC Press Web site at
http://www.crcpress.com

Printed and bound in the United States of America by
Edwards Brothers Malloy on sustainably sourced paper

To my daughters: Helya and Hannah Javaherdashti

To all who taught me and supported me over the years, especially my dear parents … Farzaneh Akvan

Contents

Authors

Reza Javaherdashti, PhD, holds a double degree in materials science and metallurgical engineering. He has more than 20 years of industrial and academic experience. In addition to various research papers and root cause analysis reports, Dr. Javaherdashti has authored several reference books on corrosion. He is an American Society of Mechanical Engineers (ASME)-approved trainer and has designed and executed many international industrial workshops. Furthermore, he has been involved in many consulting and problem-solving activities around the globe and is also a corrosion advisor to internationally renowned companies. Dr. Javherdashti is a veteran member of various well-reputed international corrosion societies such as the National Association of Corrosion Engineers (NACE).

Farzaneh Akvan has a background in physical chemistry and electrochemistry. She holds an MSc degree, and her interests include corrosion management, chemical management, and electrochemistry and its application in corrosion control. She was selected as the Best Young Researcher in Europe in 2010 for her research on cathodic disbondment (she received the award from the Gubkin Russian State University of Oil and Gas). Akvan has worked at the internationally renowned company SGS as a senior officer for inspection of chemicals and materials. In addition to industrial experience, she is also a long-standing member of the Iranian Young Researchers and Elite Club. Akvan has served as executive manager for the Iranian Corrosion Association. She is currently working as a consultant to the oil and gas industries as a senior inspector and corrosion advisor and as a lecturer on the management of the chemicals used for water treatment and corrosion. She has authored several publications on corrosion management and the link between future studies and corrosion management. Akvan is also a member of the International Electrochemistry Society.

Introduction

What is not understood—unfortunately—by many professionals is the role played by the management of corrosion in creating a "desired future" where bad incidents are minimized. Corrosion is a bad incident for which the risk—in contrast to natural disasters—is calculable and thus controllable. Figure I.1 schematically shows the relation between corrosion and a company's desired future, which could provide both economic and ecological gain.

In a corrosion-oriented futures study scheme, one has to consider corrosion as a very important factor. This is because corrosion has the ability to "tilt" an otherwise desired future of a plant (in which the corrosion costs, both economic and ecological, are minimized) into an unpleasant one where corrosion will start to cause damage and will have an adverse impact on assets.

Here, the term "corrosion" is used in its broadest sense. It can involve diverse processes ranging from corrosion under insulation to various types of corrosion under deposits,* including, but not limited to, microbiologically influenced corrosion. Hydrostatic testing is a factor that has to be seriously considered in a desired corrosion futures study scheme, because, apart from materials selection and welding, hydrotesting is the only pre-commissioning factor that can cause headaches for industry professionals when the asset (the equipment) is put into use.

Hydrostatic testing (HYD) in principle sounds quite straightforward—just fill the equipment to be hydrotested (a pipeline, a tank, a thickener, or a heat exchanger) with water, apply pressure at a level higher than faced in reality, check that there is no leak and

---

* "Under deposit corrosion" has no practical meaning to help one identify the mechanisms by which corrosion is happening. Rather, it shows "where" corrosion is happening instead of giving any clue "by which mechanism(s)" corrosion is happening.

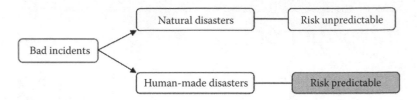

FIGURE I.1 A desired future (with minimized bad incidents) for a company would mean prosperity in terms of both economy and ecology resulting from their professional activities. Corrosion is a bad incident caused by humans, the risk of which is predictable and manageable. (From R. Javaherdashti, F. Akvan, *International Journal of Engineering Technologies and Management Research*, 2(4), 1–8, 2015.)

that the equipment's mechanical integrity survives the test, drain the water, and dry the inside. This summarizes the steps involved in any HYD. In fact, for many industries, HYD has become such a routine matter that its possible impact for inducing an unpleasant future for an asset seems to be far-fetched. Manufacturers and the contractors of the asset (a pipeline, for example) may carry out HYD without seriously reviewing the consequences, and their clients, who will then use the asset in operations, may also perform HYD themselves but not consider the required care. This is why, every now and then, a pipe is hydrotested and put into operation, then fails within a short time of commissioning, but it often goes unnoticed whether the fault was in fact due to HYD and not improper operation or unsuitable working conditions of the asset. In fact, the corrosion scenario(s) may be more complicated than first thought. It is quite possible to have competing corrosion processes and scenarios within the same asset, a concept that—to the best of our knowledge—will be introduced in this work for the first time in terms of "parallel" and "series" corrosion processes (see Section 1.1 for more details). More practically speaking, many operators and contractors do not focus sufficiently on the importance of HYD. As a result of a number of practical constraints, the water used for the test may not be as it should be, the drain/dry may not be carried out completely, and water pockets may be left in the equipment. This is in addition to the poor practice of leaving water stagnant inside equipment for long periods of time. It appears that contractors mainly focus on what they are doing at the present time; very few have a clear idea about what will happen to the equipment in the future if HYD is carried out poorly.

This book does not aim to describe how HYD has to be carried out in the sense of being a standard; rather, it aims to be an easy-to-follow guideline to assess HYD. From a long career in the industry, it has become clear that there are so-called HYD standards being applied by industries. Some of these standards are branded by very well-known standardization organizations, and others are of an in-house nature. In other words, some industries have developed their own codes of practice for conducting HYD. We have seen a good number of both general guidelines and in-house HYD standards and recommended practices, and we can certainly rank them based on how they look at the important issue of HYD "assessment." None of these, to the best knowledge of these authors, has looked at proposing a procedure to ensure that the most important corrosion scenario associated with HYD—microbial corrosion—has been addressed properly. In these recommended practices and standards one can find texts that describe how to "do" HYD, but not how to "monitor" its possible consequences. As we will see, of the four main corrosion mechanisms that may be involved in poor HYD practice, it is microbiologically influenced corrosion (MIC) that may cause the most significant problems. The relatively inadequate knowledge and practice regarding MIC in many industries and even among corrosion professionals speaks volumes about why all these standards and recommended practices are rather silent on the assessment of HYD (with regard to MIC being a very serious issue associated with HYD-induced corrosion).

This work was prepared when the authors were engaged with post-HYD risks resulting from the incorrect operation of HYD (the wrong hydrates coupled with lots of operation/corrosion management mistakes, and trial and errors). In addition, one of us (Javaherdashti) has been referenced as a referee in some international legal disputes pertaining to the failures resulting from HYD. Thus, the writing of this book arose from observing a serious gap in the literature, which for some reason had been overlooked so far.

We have kept this "Field Guide" simple, but very accurate, because we sincerely believe that truth can be expressed with simple terms. We will do our best not to introduce too many equations or even theory in this book because, as mentioned earlier, it was written from a feeling that there was an industrial need rather than as a result of a purely research-based desire. We have, however, introduced a simple way of assessing the possibility of HYD-induced MIC that can even be used to create a simple APP (we did this for a few cases ourselves). The general feature of this rather "mathematical" approach toward HYD-induced MIC is in the public domain and actually based on Frank & Morgan's principle for the assessment of factors contributing to a problem (the approach Dr. Javaherdashti used in his 2003 NACE corrosion paper to suggest an algorithm for the assessment of corrosion problems associated with a buried pipeline, which led to publishing another paper 2 years later in NACE's *Materials Performance*, describing some practical cases associated with the application of the algorithm).

This book is the first attempt to give shape to the process of assessing HYD and MIC as well as other possible associated corrosion scenarios. The lesson that must be learned from reading this text is to realize, once again, that in order to manage corrosion one has to consider various factors and their possible impacts on the overall process—an issue that can be best described by the so-called "butterfly effect": neglecting a seemingly insignificant factor could lead to huge disasters. Alternatively, this can be put simply as "Prevention is better than treatment."

We strongly suggest that our readers, especially managers, read Annex 3, which deals with corrosion knowledge management (CKM). Every corrosion problem, including HYD-related corrosion issues, can be looked at from two standpoints: an engineering viewpoint that will concentrate on controlling the risk of corrosion and a managerial viewpoint that approaches the issue from a management and cost-control angle. In Annex 3, we briefly discuss CKM as a tool for policy making regarding corrosion in general and both pre- and post-HYD corrosion risks.

We sincerely hope that our readers will find this book of some use and will apply the guidelines given here to help them assess their HYD and thus manage the risk of corrosion. We would like to finish with this extract from a poem from the great Iranian poet, Nasir Khusraw (1004–1088 CE):

درخت تو گر بار دانش بگیرد/ به زیر آوری چرخ نیلوفری را

This can be translated literally as "The more you know the more powerful you will become, so much that you can make the Heavens bow before you!"

Reza Javaherdashti and Farzaneh Akvan
Australia

CHAPTER 1

A Review of the Essentials of Corrosion Needed to Assess Hydrostatic Testing

1.1 "Parallel" and "Series" Corrosion Scenarios

As we can all remember from our school days, and especially from the study of electricity, current can flow in one of two manners depending on whether resistors have been placed one after the other or in parallel with one another. The former, called a "series arrangement," will show an overall resistance that is the sum of all the resistances, whereas in the latter, the "parallel arrangement," the overall reciprocal of the resistance is equal to the sum of the reciprocals of the individual resistances.

In dealing with corrosion processes we can use the same analogy. Sometimes, corrosive processes can take place electrochemically in a series arrangement, where each will enhance the impact of the other—a synergistic impact. On the other hand, the corrosive processes can also act in a parallel manner, seemingly independently of one another, as shown schematically in Figure 1.1.

Figure 1.1 shows that a series corrosion process may simply originate from a corrosion problem that started because of the initial conditions (the chemistry of the

1

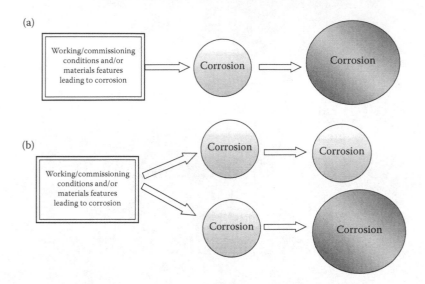

FIGURE 1.1 Corrosion processes arranged in series (a) and in parallel (b). The relative sizes reflect the severity of corrosion.

system, materials selection, hydrostatic testing [HYD], and so on). This corrosion problem may be aggravated and cause more severe corrosion problems. In a parallel corrosion scenario, however, corrosion processes can evolve independently of each other, and the severity of corrosion (measured in terms of the corrosion rate) may remain the same (upper branch) or become worse (lower branch).

Understanding the concepts of series and parallel corrosion is of vital importance in analyzing the corrosion processes that can become involved during or after HYD is completed.

1.2 Corrosion Scenarios Likely to Develop as a Result of Hydrostatic Testing (HYD) Poor Practice

Corrosion is a "thermodynamically favorable" process, which simply means that the process will happen no matter what the conditions or how perfect the prevention/control measures. Corrosion will happen anyway, and all we can do is control it to make it as slow as possible. This means that the corrosion rates must be so low that, practically speaking, corrosion can be said to have been "under control." Anything that assists the "thermodynamically favorable" process of corrosion is undesirable. Factors such as poor welding (in welded parts), the use of maltreated water,* inadequate materials selection,† and so on could enhance the likelihood of corrosion occurring.

At present, we can address four corrosion scenarios most likely associated with poor HYD (of which microbial corrosion is the most probable) [2]:

- Microbial corrosion (MIC)
- Formation of electrochemical cells (oxygen concentration counts)
- Galvanic corrosion
- Underdeposit corrosion

We will describe MIC in Chapter 2, but here we will lightly touch on three points that may seem obvious, but are, in fact, quite confusing.

- Oxygen is an integral part of any electrochemical corrosion. For this reason, oxygen scavengers are added when necessary (see Chapter 3) to exclude oxygen and thus control corrosion.
- "Dissimilar metals corrosion" is an incorrect terminology and instead of that, "Galvanic corrosion" must be preferred to be used. Galvanic corrosion can occur on the very same metal if conditions allow. A typical example is when a segment of an old pipe is replaced with a new pipe made of the same material. If the pipes have not been taken care of adequately, the new

---

* "Untreated water" means that no treatment (mainly chemical) has been applied to the water. "Undertreated water" is meant to address water that has not been treated enough to ensure that it is safe (from a corrosion point of view). Untreated and undertreated waters can be collectively termed "maltreated water."

† An example of this is using stainless steel 316L with chloride levels that are at about the same level as in drinking water but at equipment temperatures equal to or above 55°C. See, for example, ASM, *ASM Handbook*, vol. 13B, *Corrosion: Materials*, ASM, 2005.

pipe can become the anode and will corrode, despite being the same material as the rest of the pipe.

■ It is possible that the galvanic cell formed as a result of an underdeposit corrosion mechanism such as formation of underbiofilm differential aeration cells will form the required driving force for corrosion. This example alone should serve to show the complexity that may be involved in multiple corrosion reactions, setting some to each other as series and some as parallels. Another example of such is given below.

As shown in Figure 1.1, the four corrosion reactions described above can take place in series or in parallel. If, through poor corrosion management, too much deposit is formed inside a pipe, then underdeposit corrosion is inevitable. If, in addition to that, the water used for HYD is maltreated, then by increasing the likelihood of MIC we can face an unfortunate combination of both microbial and non-microbial corrosion taking place inside the pipe. Figure 1.2 shows a situation where the deposits within a pipeline have accumulated due to poor corrosion management and have formed a suitable location at which corrosion can easily proceed.*

When welding is done poorly, poor welding and post-welding treatment (PWT) will lead to the formation of platforms upon which corrosion can initiate (with or without HYD). If the conditions are made worse by applying poor HYD using maltreated water for a long period of time, the corrosion that may have happened as a result of the poor welding will be enhanced by an unfortunate combination with microbial corrosion. Figure 1.3 shows welding defects formed during HYD.

Figure 1.4 presents yet another example of how poor welding practice can result in a post-HYD failure.

---

* During further investigations it was found that these deposits were all magnetic (most probably magnetic FeS). This implies that no pigging will work thoroughly with this pipeline if it is not preceded by vigorous chemical treatment. The main aim of the treatment must be dislodging the deposits so that, upon pigging, any remaining deposits will also be removed.

(a)

(b)

FIGURE 1.2 Deposits removed after pigging through a 300 km pipe (a) and a close-up of the overall texture of the deposits (b). (Courtesy of Reza Javaherdashti.)

Therefore, it is essential to realize that what we do can enhance an already existing corrosion problem or will assist in enhancing its detrimental effect on the asset and equipment.

The outcome of poor HYD will be a loss of mechanical integrity induced by pitting (Figure 1.5). To avoid such incidents, it is essential to focus on both series and parallel corrosion mechanisms. Chapter 2 will focus on the most important corrosion mechanism induced by poor HYD—microbial corrosion.

Although HYD seems to be a "continuous" routine process, in the sense that it involves seemingly simple tasks (i.e., "simple" when compared to tasks such as

FIGURE 1.3 Example of a failed line pipe manufactured by low-frequency electric resistance welding (LF-ERW) after HYD. (From J.P. Sinha, C.P. Varghese, *Assessment of seam integrity of an aging petroleum pipeline constructed with low frequency ERW line pipes*, 6th Pipeline Technology Conference 2011, Germany, 2011, http://www.pipeline-conference.com/abstracts/assessment-seam-integrity-aging-petroleum-pipeline-constructed-low-frequency-erw-line. With permission.)

FIGURE 1.4 Brittle fracture during a pipeline hydrotest due to a discontinuity in a submerged arc welding (SAW) longitudinal weld. (From Massimo Benedetto. With permission.)

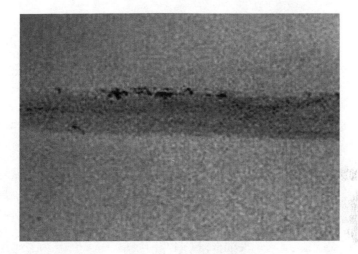

FIGURE 1.5 Pitting resulting from poor HYD (material: carbon steel; HYD medium: fresh water). (Adapted from R. Javaherdashti, *Corrosion Management*, January–February, 2009.)

pigging subsea pipelines or even continuous casting), in fact it is a rather intermittent issue. The material from which the pipe is made is manufactured in a shop, but the HYD medium is not performed there. The HYD medium and its treatment are sourced from another contractor. We then have tasks that are seemingly independent of each other: water flooding in a pipe and drain/dry can be regarded as tasks that can be classified as one single module. These can all be summed to let us classify them in accordance with their relative contribution to the corrosion risk that may result from hydrostatic testing.

1.3 Classification of HYD Implementations and Associated Risks

The risks associated with HYD can be categorized into two groups:

■ Group 1: Pre-HYD risks
■ Group 2: Post-HYD risks

Group 1 risks can include the following.

▪ *HYD medium*: Although, from a practical point of view, it is nearly always natural water from ponds, seawater, rivers, or wells that is used as the HYD medium, there are alternatives, such as demineralized water (DW) or high-purity steam condensate (HPSC). The pros and cons of these have been discussed elsewhere [5]. Based on the water conditions, we can divide all HYD processes into two groups:
   ▪ Wrong HYD, where the water has had bad or no chemical treatment (that is, maltreated water is used for HYD).
   ▪ Inadequate HYD, where either draining or drying has not been done thoroughly or has been done so poorly that water is left inside the asset.
▪ *Materials*: Carbon steel is available for most assets like pipelines, and, based on factors such as availability of material, it can be upgraded to corrosion resistant alloys, such as stainless steel 316/316L.
▪ *Manufacturing/preparation*: Has the pipe been welded and, if so, has post-welding treatments (PWT) been applied correctly?

Group 2 risks can include the following.

▪ Leaving water stagnant in the equipment for a long period of time after HYD is completed. An example could be wet layup with maltreated water.
▪ Inadequate draining (leaving water pockets behind).
▪ Inadequate drying (leaving water pockets behind).

It must be noted that although we have classified these risks into pre- and post-HYD, they are quite likely to cause both series or parallel corrosion failures. If the material is selected inadequately and this poor choice is accompanied by introducing poor quality water into a pipe whose welding cannot be trusted and the drain/dry is also not perfect, the corrosion scenarios thus triggered could have a series effect on each other despite the fact that some of these factors were in place before HYD (e.g., materials selection, welding, and choice of HYD water) and some afterwards (inefficient drain/dry).

Another rather important issue that must be taken into account here is that the treatment policy for any corrosion resulting from HYD can have a tremendous effect on post-HYD costs. If HYD is not carried out carefully and, subsequently, there are localized leakages, then the owner/operator of the line will undoubtedly initiate quite unpleasant legal disputes against the consultant and contractor. These legal issues will certainly delay commissioning, and they will also have a huge adverse impact on the economy of the job. It is thus advisable for both consultants and HYD contractors to be aware of these two groups of risk and perhaps to do some calculations to see whether the costs associated with the prevention of post-HYD corrosion are more significant than the costs they could easily afford for pre-HYD and during HYD operations. The experience of the authors is that smart contractors will do a job that adheres to the principle of "Prevention is better than treatment."

CHAPTER 2

Microbiology of Corrosion, Microbiologically Influenced Corrosion (MIC), and Its Role in HYD

M icrobiologically influenced corrosion, also known as microbial corrosion, has three features [6]:

- It is an electrochemical process.
- Microorganisms are capable of affecting the extent, severity, and course of corrosion.
- In addition to the presence of microorganisms, nutrients and water must also be present to initiate MIC.

Figure 2.1 shows the factors that can contribute to MIC, including HYD.

In contrast to what many—including some corrosion professionals—may think, MIC is not limited to just a few bacteria. In fact, it is not only corrosion-related bacteria (CRB) such as sulfate-reducing bacteria (SRB) that affect the severity of corrosion, but also their "cousins," including the sulfate-reducing archaea (SRA). Table 2.1 presents a classification of some CRB and archaea (extracted

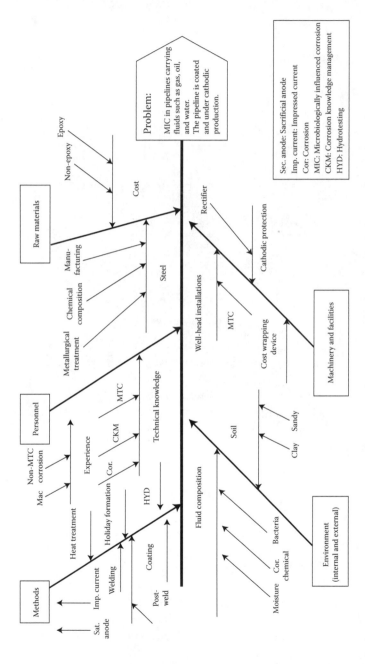

FIGURE 2.1 Factors that contribute to rendering a system vulnerable to microbial corrosion (including HYD) and how to control them. (From R. Javaherdashti, *Microbiologically Influenced Corrosion—An Engineering Insight*, 2nd edition, Springer, UK, 2017.)

TABLE 2.1 A Selection of Corrosion-Related Bacteria and Archaea Normally Found in Water (e.g., Seawater)

General heterotrophic bacteria/ archaea	Thiosulfate-oxidizing bacteria (TOB)	Acetogenic bacteria	Slime-forming bacteria
Acid-producing bacteria/ archaea	Methanogens	Nitrate-reducing bacteria/ archaea	Sulfur-oxidizing bacteria/ archaea
Sulfate-reducing bacteria/ archaea	Nitrite-reducing and sulfur-oxidizing bacteria (NR-SOB)	Thiosulfate-reducing bacteria (TRB)	Thiosulfate-reducing archaea (TRA)

from Al-Abbas et al. [8]). It should be noted that in this classification we have not included bacteria such as magnetotactic bacteria, which may make significant contributions to MIC, as theorized in Javaherdashti [7].

The bacterial groups indicated in Table 2.1 provide an idea of how complicated MIC can be. One consideration with regard to MIC, especially HYD, is to differentiate between "planktonic" bacteria (freely swimming) and "sessile" bacteria (motionless, also termed "biofilm").

It may be a good idea to examine at least five groups of these bacteria that could be linked to corrosion: sulfate-reducing bacteria (SRB), iron-reducing bacteria (IRB), sulfur-oxidizing bacteria (SOB), iron-oxidizing bacteria, (IOB) and Clostridia. The reason we have highlighted these five groups of bacteria is not that they are the "most important" corrosion-related bacteria, but because if any of these bacteria are present in corrosion products after HYD or in the HYD water, this will be a serious indication that there is a high risk of MIC and thus is not to be taken lightly. More detailed information about these bacteria is provided elsewhere [6,7].

Before discussing the CRB, we have to say a few words about an important aspect of microbes that can have a tremendous impact on the microbiology of corrosion post-HYD: the classification of bacteria based on their oxygen demand. Based on their oxygen demand regime, bacteria can be divided into aerobic (oxygen-demanding) and anaerobic (no oxygen) bacteria.* Although this classification is not particularly precise, it can serve our purpose from an engineering point of view, especially with regard to HYD and its MIC risks.

Aerobic bacteria are rather simple (they need oxygen), but when it comes to anaerobes, life becomes complicated. In almost the same way as in electrochemistry, where one can define the anode as the electron giver and the cathode as the electron taker, when it comes to biological species we can talk about electron acceptors and electron donors. The importance of this terminology as applied to anaerobic bacteria lies in "where" the electron acceptors are located; one can have anaerobic bacteria that can do "anaerobic respiration" (a seemingly oxymoronic term) and fermentation. Let's define these terminologies in a way that even an engineer can understand!

The following are the main features of anaerobic respiration. Electrons are released either by organic carbon "oxidation" (e.g., via lactate under laboratory conditions) or by hydrogen "oxidation" (by methanogens, some SRB, for example), while electrons are absorbed by external electron acceptors such as, but not limited to, sulfate, thiosulfate, sulfite, sulfur nitrate, nitrite, and CO_2. This would mean that one can find SRB (the bacteria that reduce sulfate to sulfide, see below) in environments where sulfate may not be available as the bacteria can feed on nitrate, for example. Another consideration is that if SRB are present in an already contaminated post-HYD situation and one applies anodic inhibitors that may contain nitrates,

---

* To simplify, we have deliberately neglected more detailed, and for that matter more precise, classifications, those that also include facultative anaerobes, microaerophiles, capnophiles, and so on.

although it might affect non-MIC corrosion, it will allow the SRB to grow and become more aggressive.

Another option for anaerobic bacteria is fermentation. The following three characteristics are critical:

- There are no external electron acceptors.
- Fermentative microorganisms (such as acid-producing bacteria [APB] or SRB) produce their own electron acceptors (organic carbon from a carbon source).
- The by-products of fermentation are organic acids (e.g., acetic acid) and alcohol.

It may be seen, therefore, that if the required level of care is not given to pre- and post-HYD risks, the bacteria that may have contaminated the equipment could later help intensify corrosion in the form of MIC.

Many CRB have been identified, and we mention just a few of them in Table 2.1. However, some of the most significant CRB in terms of their contributions to corrosion can be briefly introduced as follows.

SRB are exotic microorganisms in the sense that in a planet where the availability of oxygen is an essential requirement, these bacteria are anaerobic; in other words, they do not require oxygen. The impact of SRB on the corrosion of metals was discovered toward the end of the nineteenth century. The finding that bacteria could cause corrosion in addition to causing health problems, coupled with the fact that oxygen is in fact like a poison to them, has generated great interest among researchers (and non-researchers as well). There are even theories that claim SRB are actually guests from outer space. In any case, their impact on corrosion is that they reduce sulfates into sulfides. If sulfide can find metallic ions such as ferrous ions, they combine to produce the black rust of iron sulfide. There are theories suggesting that the iron sulfide thus produced will create a galvanic couple with the underlying steel substrate, and the iron sulfide, as the cathode, will severely corrode the underlying steel anode. The best theory to describe MIC is perhaps "electrical microbiologically influenced corrosion" (EMIC). However, if the sulfide cannot find metallic ions, it will combine with hydrogen to produce hydrogen sulfide. The gas thus produced is itself corrosive, and if it comes into contact with water it will produce low pH conditions that in turn will accelerate corrosion. Typical recorded corrosion rates are in the range of 1.8 mm/yr or even higher (in a particular case for a subsea pipeline in 1989, the corrosion rate due to SRB was estimated as 10 mm/yr). If Carbon:Nitrogen (C:N) < 10, SRB growth can be very noticeable. In addition, SRB can tolerate pH levels between 5 and 9.5 and can also grow at temperatures between 4°C and 110°C.

IRB reduce ferric to ferrous ions and thus lead to a greater likelihood of anodic depolarization. The way they contribute to corrosion can be explained in simple terms: ferric ions (Fe^{3+}) are insoluble and thus "good" from a corrosion-protection point of view, because, when they form, they create compounds that can cover the surface of the metals like a coating, and this coating will not allow the anode and cathode to see each other or the anodic and cathodic sites to see the electrolyte. The end result is that corrosion will be highly controlled. However, when IRB are present, they reduce the "good" ferric ions to "bad" ferrous ions. Ferrous ions are not protective and are "soluble." Therefore, in the presence of IRB, the good coating is dissolved and the metallic substrate is presented to the corrosive electrolyte over and over again. The outcome will be to enhance corrosion.

SOB are aerobic. SOB are capable of producing sulfuric acid with a pH of around 1 by one of the following reactions:

$$H_2S + 2O_2 \rightarrow H_2SO_4$$

or

$$2S + 3O_2 + 2H_2O \rightarrow 2H_2SO_4$$

It is believed that SOB such as *Acidithiobacillus thiooxidans* form "biological" sulfur that is hydrophilic (in contrast to non-biological sulfur). This is thought to be due to the formation of a coating (phosphatidylglycerol) around the sulfur, which allows the biological sulfur to wet and disperse in water. Due to their ability to produce a highly acidic metabolite, SOB are also used in bioleaching processes where the metal inside the ore is extracted biologically.

Clostridia have been known since the 1880s. They are not only important in corrosion, but also in medicine, as they are capable of inducing gangrene. Clostridia are spore-forming anaerobes that need no sulfate. They are also able to produce hydrogen and organic acid. Acids produced by bacteria accelerate corrosion by dissolving oxides (the passive film) from a metal surface and accelerating the cathodic reaction rate. The produced hydrogen can lead to hydrogen-induced cracking (HIC). Some Clostridia have been reported to have the ability to generate hydrogen sulfide gas or, like *Clostridium butyricum* (a butyric-acid-producing species), be capable of reducing iron as well. This means that there are three important contributions Clostridia make to MIC:

- Enhance anodic reactions by producing acids
- Facilitate HIC via hydrogen generation
- Enable the constant availability of a freshly corroding steel surface as a result of ferric iron reduction

Based on the microbial communities in HYD water, there may be one or more of these bacterial types.

2.1 Freely Floating and Motionless Bacteria and Their Contribution to MIC

The water used for HYD is normally provided from natural sources of water such as seawater, rivers, and ponds. These waters normally contain bacteria, among which CRB are quite common. Each liter of seawater contains one cell of SRB [2]. These bacteria are normally in a "planktonic" state (i.e., freely swimming). When the water comes into contact with a surface and the food required by bacteria falls onto the surface, the bacteria "sense" to find the food and establish themselves on these surfaces. In this state, the bacteria are no longer swimming, but are tied onto the surface. This bacterial state is termed "sessile" or "motionless." It is during this stage that "biofilms"* are formed. The majority of corrosion in the form of MIC arises from biofilms, which is why their contribution to HYD must be taken seriously. Figure 2.2 summarizes, schematically, the steps of biofilm formation.

The importance of biofilm formation for HYD is that if the HYD water is maltreated and drain/dry is not carried

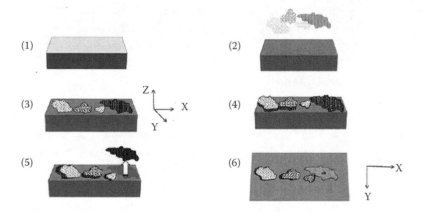

FIGURE 2.2 Six stages of biofilm formation: (1) bare metal surface, (2) bacteria in planktonic state, (3) bacteria have settled on the surface, (4) the film thus formed starts to thicken, (5) the bacteria die off due to lack of oxygen and food, and (6) due to factors such as the formation of differential aeration counts, pitting under the biofilms is facilitated. (Adapted from R. Javaherdashti, in H. Kanematsu and D.M. Barry (eds.), *Biofilm and Materials Science*, Springer, Switzerland, 2015.)

---

* In fact, biofilms are neither totally biological nor a film.

out properly, the likelihood of biofilm formation and corrosion being induced after HYD will be quite high. This is very important. During normal periods of HYD (between 4 and 8 hours, see Chapter 3), it is quite unlikely that biofilms will develop,* but in the post-HYD period, the asset may be at risk of biofilm formation. In raw seawater (which is rich in organic matter), it normally takes more than a week for a biofilm to mature, but in filtered seawater this extends to more than a month [2]. It follows, then, that if seawater, as the HYD medium, is left stagnant inside an asset (a pipe, for instance) for more than a month, mature biofilm will be generated and is very likely to lead to corrosion in the form of MIC.

2.2 MIC: A General Scheme

When equipment is tested and put into service, or even sometimes before it is put into service, HYD-triggered MIC may occur. In other words, while MIC in other circumstances may be regarded as a problem arising during commissioning and equipment use, MIC produced as a consequence of wrong/inadequate HYD can be regarded as a pre-commissioning problem. It is thus important to have a general understanding of MIC.

There are three important stages to consider in relation to MIC (Figure 2.3):

- Recognition
- Treatment
- Monitoring

We will briefly explain these based on the codes given in Figure 2.3.

---

* In any post-HYD operation (such as pigging), if complete removal of the biofilm and debris is not done, the asset will be highly susceptible to MIC due to regrowth of the biofilm. (See, for example, R. Javaherdashti, Circumstances where biocides may be ineffective, *Materials Performance*, 47(11), 60–63, 2008.)

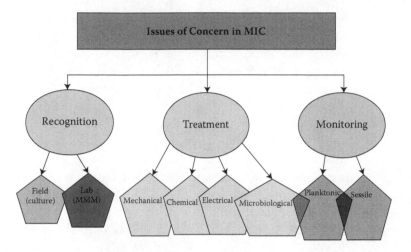

FIGURE 2.3 Three aspects of MIC. (MMM, molecular microbiology method.)

What is meant by recognition is how we can recognize from the morphology (appearance) of the corrosion what type of bacteria* are responsible for the MIC-induced failure. The recognition task takes place only after all the evidence has suggested that the corrosion is indeed MIC. In the particular case of HYD, there is a chance of the corrosion being non-microbial in origin. So, after we have collected enough evidence to believe that the corrosion case is MIC, we need to specify the type of bacteria involved.

There are two methods used to achieve bacterial recognition. The first is a quick, relatively inexpensive, but less accurate response, and the second is relatively expensive and labor-intensive, but achieves much more precise results. The former are "field" methods and the latter "laboratory" methods.

Field methods are based on the principle of "culture-dependent methods," such as serial dilution or "most

---

* An important difference between MIC cases and medical cases, both related to bacteria, is that in medical cases it is always one type of bacteria that causes the (health) problem, but in industrial cases it is usually more than one type of bacteria that contributes to corrosion. In the latter case, there can be a "sulpureta" formed where a bacterial species will facilitate the growth conditions for another species.

probable number" (MPN). Laboratory methods can also include serial dilution and MPN, but they mainly focus on the so-called "culture-independent methods," also known as "molecular microbiology methods" (MMM). Examples of culture-independent methods include the following:

- Denaturing gradient gel electrophoresis (DGGE)
- Fluorescence in situ hybridization (FISH)
- Microbial diagnostic microarray (MDM)
- Metabolic polymerase chain reaction (PCR)
- Denaturing activity measurement techniques such as adenosine triphosphate (ATP) measurement
- Next-generation sequencing (NGS)

The pros and cons of culturing for a corrosion engineering approach can be summarized as follows. Culturing can be used not only to analyze pure cultures (either in terms of morphology but also in terms of "action"), but also to maintain such cultures as a "back-up" for further investigations. Culturing is a "pre-defined" approach, in that when certain bacteria are allowed to grow in a given medium, it is always possible to "miss" others. In other words, many bacterial species with known and unknown effects on corrosion may be present in a corroded sample, but, with culturing, only a certain number of these microorganisms will be revealed (e.g., SRB and IRB). The best approach is to verify the results of culturing methods using other methods. This, although adding to the cost, will result in more accurate results, so that a more realistic mitigation approach can be designed and applied.

Culture-independent methods also have their pros and cons:

- Quantitative PCR (qPCR) measures the number of microorganisms in a sample by quantifying their genetic material; for example, there is only one gene (*dsr*) corresponding to SRB (one gene/ one SRB bacterium), and one gene (*mcr*) present in methanogens.

■ qPCR can measure the total number of bacteria/ archaea, the numbers of three groups of methanogens, SRB and SRA.

■ The qPCR method can be designed to be highly selective towards particular genes and microorganisms in a system. The user should first characterize the microorganisms that are present.

■ DGGE cannot detect microorganisms such as methanogenic archaea.

■ NGS can differentiate between bacteria and archaea.

The "ATP measurement" makes use of the reaction in which ATP, in the presence of the enzyme luciferase, catalyzes the oxidation of luciferin. As a result, ATP yields adenosine monophosphate (AMP), pyrophosphate, and light. The intensity of light is proportional to the amount of ATP present, and thus the number of bacteria.

$$\text{ATP} + \text{Oxygen} \xrightarrow[\text{Luciferase}]{\text{Mg}^{++}} \text{AMP} + \text{PPi} + \text{Oxyluciferin}$$
$$+ \text{Luciferin} \qquad\qquad + \text{LIGHT}$$

Table 2.2 summarizes pros and cons of ATP G1. To overcome the shortcomings of ATP G1, second generation of ATP methods (ATP G2) has been developed. ATP G2 can quantify total active micro-organisms in any water, wastewater, organic fluid from a wide range of industries. This can be done by measuring total ATP (tATP) or cellular ATP (cATP). Deposit and surface analysis (DSA)-tATP can measure sessile biomass on a surface whereas quench-gone organic modified (QGOM) is classified as a cATP method applicable to water, organic matter, and finished product samples.

TABLE 2.2   Pros and Cons of ATPG1

Pros	Cons
This method has been used to estimate the relative total number of bacteria in environments where "non-bacteria" ATP is rare. This method can be carried out in less than an hour.	ATP G1 cannot distinguish between ATP extracted from bacteria and other organic debris in the sample, and is vulnerable to interference. The amount of ATP is not predictable in SRB and some other common environmental bacteria, giving rise to *very approximate values for total bacteria count.*

It is essential for those who want to assess post-HYD risks to know these priorities in order to assess the hydrotesting procedure.

All of these methods are widely available and, with appropriate instructions from the engineer as the knowledgeable client, an experienced microbiologist can use these and other methods to identify the bacterial communities that may have been formed during post-HYD stages such as wet layup or even during wet parking, when the conditions are favorable for the formation of biofilms, initially harmless and inoffensive microbial communities, which later will turn into aggressive microbial biofilms and accelerate corrosion.

As most corrosion-related issues related to HYD occur after the hydrotest, it is advisable to apply culture-independent methods and not just rely on culture methods. Culture methods have the intrinsic disadvantage that they normally measure the number of planktonic bacteria in water. However, the majority of corrosion arises not from planktonic bacteria, but from sessile bacteria. If culture-based methods such as rapid absence/presence tests are applied to the water that is to be used for HYD, they may be advantageous in the sense that they can give a very rough idea about the MIC potential risk posed by the water.

Most of the time it is water from natural sources (such as seawater) that is used for HYD, and the only treatment for Group 1 risks (the pre-HYD) might be the application of biocides (to manage MIC) and inhibitors (to manage electrochemical, non-MIC corrosion). However, if Group 2 risks (post-HYD) occur, then, based on the condition of the equipment, either physical or chemical measures can be applied to mitigate MIC. It is also advisable that, when appropriate, a combination of these methods should be used, for example, damaging the biofilm and applying a single- or dual-regime biocide [7].

CHAPTER 3

Assessment of HYD

For those who have been engaged in performing HYD there is nothing too confusing: the standards are known, the procedure has been known for over 20 years, so no problem! The reality, unfortunately, is that "performing" a task is very different from "assessing" if it is being done correctly. HYD is a good example of this.

Hydrostatic testing is a leak+ strength test. In other words, it has two components, and for that reason, whatever is recommended (and done) must consider these two factors simultaneously. Accordingly, this chapter does not dictate "how" to do HYD but rather provides a "guide" as to how to do it to minimize unwanted post-HYD trauma—in other words, corrosion and particularly MIC.

3.1 Factors Important in HYD

Among the various factors that can be significant in affecting HYD quality, the following six factors are of importance[*]:

- HYD water and its source
- Pipe material

---

[*] It goes without saying, here, that our fundamental assumptions are that the HYD fluid is water and the system on which HYD is being applied is a pipeline. However, as most of the time the HYD medium is water, the six factors mentioned will still be valid for any system other than pipelines.

- Intended application
- Chemical treatment applied to HYD water
- Wet parking duration (the time during which water remains inside the pipe after HYD has been tested and is finalized)
- Discharge

We will briefly concentrate on four of these which are the most important regarding the context of this book.

3.1.1 HYD Water and Its Source

3.1.1.1 Where to Get HYD Water

To contribute to the economy of an operation, the water used for HYD is usually the water that is easiest to reach, but this is risky if the water has not been filtered or chemically treated. This is because this "raw" water could have a great risk of harboring and nourishing CRB that, could affect the integrity of the pipe by inducing corrosion later. Water used for HYD can be ranked from demineralized (most suitable) to brackish (least suitable), but with clean seawater (mentioned below) placed above river or lake water in terms of suitability for HYD [2].

The following must be considered when selecting HYD water sources:

- If seawater is to be used it must be "clean and clear" and taken from the "safe zone" of seawater. This "safe zone" is the clean and clear water taken from the seawater layer that is 15 m above the seabed (to minimize the likelihood of there being too much sand and sediments and thus a higher total dissolved solids [TDS]) and 15 m below the seawater surface (to minimize the likelihood of getting too much nutrient content, which can later be used by CRB in HYD water) [2].
- It is always possible to use other test media instead of water. Table 3.1 summarizes the pros and cons of two of such "alternatives," but the cost may never

TABLE 3.1 Some Pros and Cons of Possible Alternatives to HYD Water

Alternative	Pros	Cons
Demineralized water (DW)	Provides draining and drying of the system at the earliest opportunity after hydrotest Disposal is usually not a problem	Costly Difficult task of drying a large process system after testing
High-purity steam condensate (HPSC)	Provides draining and drying of the system at the earliest opportunity after hydrotest Depending on the chemicals present in the condensate, disposal may not be a problem	Costly Practical problems with finding a chemically clean, large steam system for testing a large process system More problematic than the DW approach

Source: R. Javaherdashti, Enhancing effects of hydrotesting on microbiologically influenced corrosion, *Materials Performance* (MP), 42(5), 40–43, 2003.

be comparable to using "ordinary natural-sourced" water. It is thus important to not only look at the possible pros of an alternative to HYD water, but also the cost.

■ As one of the most important factors in HYD is the material of the structure (e.g., the pipeline), the compatibility of the HYD water and the material must be considered carefully (see Section 3.1.2).

3.1.1.2 Microbiological Assessment of HYD Water

MIC is the result of CRB. It is therefore essential to know whether the HYD water harbors biological activity that will later induce corrosion should the water remain in the pipe (as frequently occurs in real-life situations). There are two ways to detect CRB in HYD water: by laboratory and field tests (Figure 3.1).

The details of culture-dependent and -independent tests have been discussed elsewhere [7]. The main thing to recall is that, for many reasons (ease of performance, transportation, cost, and so on), what is performed in the

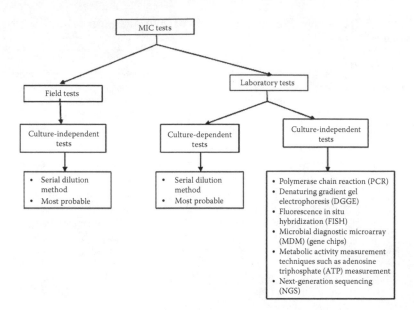

FIGURE 3.1 Two major categories of MIC detection methods, with some examples of each subcategory.

field is not a laboratory test but a field test, as the name implies. As shown in Figure 3.1, field tests are essentially culture-dependent methods. All these methods therefore have serious limitations, which can be summarized as follows: they are capable of revealing only a very small portion of the microorganisms (those that are cultivable); they may be too time-consuming; the culturing environment can be different from the natural environment of the microorganism; they are selective (i.e., they grow only "pre-selected" microorganisms) [7]. Therefore, although the use of culture-dependent field tests is inevitable for HYD, the following must be considered:

■ One should know exactly what types of bacteria are to be tested. There is no field test that can reveal (culture) all CRB. As a suggestion, the following CRB must be taken into account when testing

HYD water: sulfate-reducing bacteria (SRB), sulfur-oxidizing bacteria (SOB), iron-oxidizing bacteria (IOB), iron-reducing bacteria (IRB), and Clostridia. IRB and IOB can be collectively referred to as "iron bacteria," while SOB and Clostridia can be collectively addressed as acid-producing bacteria (APB).

- Inoculation must be done at the sampling point. The longer it takes for the HYD water/solid sediments to reach the laboratory, the greater the risk of serious errors [7].

- The freshness and validity of the test kit itself should be checked. Discard expired tests kits. Field "rapid test" kits are always subject to modifications and changes, so it is a must to check with a manufacturer whether certain kits have been discontinued.

- The kit should be user-friendly in terms of application and transportation. The test kits must have been designed in such a way that they can be used with minimum supervision. Test kits that use solid (preferably powdered form) culture media are preferred to those that use liquid media because of transportation issues. Because manufacturing CRB rapid check field test kits is a commercially profitable business, there are many such kits in the marketplace. Some even require the use of syringes to introduce the water sample into the culture medium. Using such test kits may create serious problems in transportation and application.

3.1.2 Pipe Material

Under certain circumstances it may become necessary to switch from the usual material choice (carbon steel) to another material (stainless steel family). Of these, austenitic stainless steels (SS304 and SS316) may be the first choice.

This material selection may be based on various considerations. In one case, the pipe manufacturer could not provide a specially lined carbon steel pipe, so the operator had to opt for SS316L. In such cases, one must be very careful with the chloride content and operating temperature (or, even better, the actual HYD water temperature during both HYD and post-HYD operations such as wet parking). It has been reported [10] that below 30°C, stainless steel 316L can offer good pitting resistance if the chloride content is 1000 mg/L. When using seawater (on average containing 35 g of salt per liter), however the chloride content-temperature combination could be detrimental if HYD is being carried out where the HYD water is likely to reach temperatures above 30°C. One thus has to be careful about three factors: the type of HYD water, the water temperature during and after HYD, and the material used for the pipe/structure.

If the pipe is constructed from carbon steel, the addition of oxygen scavengers may not be required. This is an issue that is important from a practical point of view, and we will deal with it in Section 3.1.4.

3.1.3 Intended Use

Whether the pipe is to be used for an onshore application or offshore, and whether it will be used to transport gas or liquid, are all important issues. For example, if the pipe is to be used for offshore applications it will require increased relative thickness compared with an onshore pipe. This will, in turn, have a significant impact on the pressures that can be applied to the pipe to measure leakage and strength (which are the ultimate aims of any hydrostatic test). Also, whether the pipe will carry gas or liquid in its future service will have an impact on the duration of the HYD.

3.1.4 Chemical Treatment Applied to HYD Water

As HYD water is generally a raw water in the sense that it comes from natural sources and has not been specifically treated for the purpose of controlling its corrosivity (both MIC and non-MIC), it is essential to use chemicals to condition it. There are three chemicals that can be added into HYD water:

- Oxygen scavengers
- Corrosion inhibitors
- Biocides

In practice, some contractors do not apply any of the above chemicals to HYD water on the basis of cost and operation time. Operators must be aware that if they do not require their contractors to carry out the above treatments, the future safe operation of the pipe will be at risk of leakage and perhaps failure.

3.1.4.1 Oxygen Scavengers

Oxygen reduction is an inevitable part of any electrochemical corrosion process that takes place under aerobic conditions, so finding a way to exclude oxygen is a key factor in controlling corrosion. For this reason, certain chemicals (such as bisulfites) are added into HYD water. Examples of oxygen scavengers include ammonium, amines, and sodium biosulfite. In a typical dosage of oxygen scavenger, for each part of oxygen present, eight to ten parts of biosulfite will be necessary to exclude cathodic oxygen. There is, however, an important issue to consider regarding the application of oxygen scavengers when the pipe is made of carbon steel, in that there is the possibility of an interaction between oxygen and the pipe wall. This is described in Kochelek [11].

When a pipe is filled with water, this water replaces the air inside the pipe. On increasing the volume of water, trapped air will remain in the water. One can imagine that there are two sources of oxygen in such a closed system: oxygen from the trapped air inside the pipe and oxygen already present in the water. First, the dissolved oxygen starts to react with the pipe walls and corrosion starts. This process will continue until the oxygen dissolved in the water falls below its saturation limit, which is about

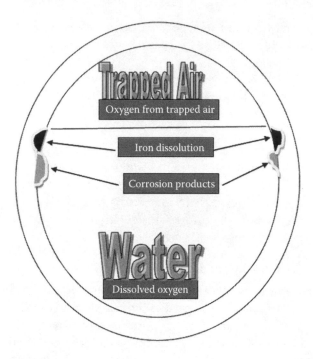

FIGURE 3.2 Possible iron–oxygen reactions within a carbon steel pipe filled with water.

10 ppm (parts per million) at room temperature and pressures [12]. Oxygen will then be supplied from the trapped air within the pipe until all of the oxygen is consumed. The procedure is shown schematically in Figure 3.2.

Oxygen from trapped air: When a pipe is filled with water, the air inside the pipe is replaced with water and gradually trapped and compressed. This air has oxygen in it—1000 gallons of air that originally filled the pipe is compressed and trapped in the piping after filling with water. This contains 210 gallons of oxygen.

Dissolved oxygen: The water already has some oxygen in it as dissolved oxygen. A volume of 1000 gallons of water completely saturated with dissolved oxygen to 40 ppm (0.004%) contains 0.04 gallons of oxygen (assuming 150 psig pressure in the pipe).

The cathodic reactions necessary to sustain corrosion (i.e., oxygen reduction) will be supplied by these two sources. The anodic reaction will be the dissolution of iron, manifested as pitting, and the cathodic reaction will be demonstrated as corrosion products or rust. (Data regarding the amount of oxygen in the water and air are directly taken from Kochelek [11]).

The same phenomenon occurs when HYD water is left within a carbon steel pipe. Depending on the water used for hydrotesting (seawater or fresh water, temperature, thickness of the pipe), it normally takes between 18 and 48 hours for the dissolved oxygen to drop from parts per million (ppm) levels to parts per billion (ppb) [13]. This "oxygen consumption" phenomenon has a great impact on the addition of oxygen scavengers into the HYD water in the sense that many operators may actually choose not to apply oxygen scavengers during HYD. If after HYD, however, the layup is to be a wet layup (i.e., filling the pipe with water and preserving it), it may be a good idea to add oxygen scavengers, as long as the right regime of biocide application is implemented and there is no concern that, in the absence of oxygen, anaerobic CRB will proliferate and induce corrosion.

Like any other chemicals, when oxygen scavengers are combined with other chemicals, the resulting effect may be antagonistic, meaning that either one or both chemicals may show an altered/reduced chemical impact. One example of this is the use of glutaraldehyde and tetrakis hydroxymethyl phosphonium sulfate (THPS) (both non-oxidizing biocides) with a bisulfite oxygen scavenger. Both biocides rapidly degrade when exposed to oxygen, so oxygen must be excluded by adding an oxygen scavenger. However, to prevent antagonistic effects, it is recommended that first you add the oxygen scavenger, wait 5 minutes, and then add the biocide [2].

3.1.4.2 Corrosion Inhibitor

Corrosion inhibitors are chemicals that can alter the severity of corrosion by reducing its rate via various mechanisms that can affect anodic reactions, cathodic reactions, or both. Details of the way inhibitors perform have been discussed elsewhere [7].

In terms of corrosion inhibitors, the following factors should be considered:

- Generally, corrosion inhibitors are not the same as biocides. Corrosion inhibitors are mainly designed to control non-MIC corrosion, and biocides to control MIC by killing the CRB involved. It is possible to find some inhibitors that may also have biocide effects (e.g., alcohols) or biocides whose biocidal impact is more electrochemical than biological (e.g., hydrogen peroxide [7]), but these chemicals must not be addressed interchangeably.
- If the HYD water is suspected of harboring active CRB, using anodic inhibitors such as nitrates is *not* recommended. In HYD systems that are already contaminated with CRB, care must be taken not to use chemicals that are either biodegradable or contain organic carbon. In both cases, the added chemical may be used as a nutrient. We have witnessed such cases: because operators were not concerned/aware of MIC they applied a chemical as a corrosion inhibitor, even though its MSDS (material safety data sheet) clearly stated that the inhibitor had an organic composition. After applying the chemical, the number of planktonic SRB increased, mainly due to inappropriate chemical management and selection.

3.1.4.3 Biocides

Biocides are chemicals that can kill bacteria. One way of classifying biocides is by categorizing them into "oxidizing" and "non-oxidizing" biocides. Chlorine and ozone are examples of oxidizing biocides, and THPS and quaternary phosphonium salts (quats) are examples of non-oxidizing biocides. Both categories have their pros and cons, which are discussed elsewhere in detail [7]. It is important to be aware of the limitations that biocides may have; for instance, the non-oxidizing biocide isothiazolone is quite effective against SRB and *Pseudomonas*. However, in systems containing sulfide, bisulfite, mercaptans, organic acids (cysteine), and amines, this biocide will lose its effectiveness [14]. On the other hand, THPS is a biocide that also shows chemical effects such as dissolving iron sulfides [15].

Biocides must have the following features:

- *Broad spectrum*. The biocide must be able to kill as many and as diverse bacteria/algae/fungi as possible. For example, metrodinazole is effective only against SRB, and carbamates are effective against both SRB and spore formers, so it is prudent to prefer the latter over the former, other things being equal. The only exception to this is where, on continuous monitoring of the system, it becomes evident that there is only one type of CRB causing problems.

- *Economically compatible*. This not only refers to the overall cost of the chemical, but also its cost per dosage. This means that the biocide must be effective at low dosages/concentrations. Using gaseous chlorine can be more economical than using chlorinating compounds, as the former would require smaller quantities than the latter, other things being equal.

- *Ecologically compatible*. This means that the biocide must not be toxic, especially to humans. If one has the choice of ozone or acrolein under equal circumstances, choosing ozone will be the wise option.

- *Cannot be antagonistic with other chemicals*. In any system, like a pipe under HYD, it is expected that there will be a cocktail of chemicals of which biocides are just a part. This cocktail must not have corrosive effects on the material itself.

Finding the optimum concentration of a biocide is a task to be made based on a decision matrix that itself has been designed according to the four features given above. The broad-spectrum nature of the biocide rests on first knowing what is inside the system. Thus, it is necessary to have an overall understanding of the bacterial community in the system, then we can use some measures that, although from a hardcore scientific point of view may not be discussion-free, from a workshop point of view can help the field engineer. An example of such "operator guidelines" is provided in Olabisi et al. [16]. In such practices, certain population density numbers for both planktonic and sessile bacteria (SRB, general aerobic bacteria, and general anaerobic bacteria) are set as a target, and the aim of the operator is to reach these numbers. As said, these numbers are just indicators for the field operators and have no value per se; for example, when the number of SRB (either planktonic or sessile) is set at a certain level, this ignores the fact that there is no link between the number of SRB and the corrosion rate [7]. An example of the procedures to test chemicals (including biocides) for HYD is given in Prasad [17], which states how biocides can be tested in pH-adjusted media against SRB. The same methodology can be extended to other CRB based on the fact that we know what CRB are in fact present in the HYD system; based on that, the required tests are designed and applied.

It is important to bear in mind, the following points:

■ Biocides have to be applied, especially if the HYD water is to remain in the pipe. As the main concern with MIC is biofilm formation, all efforts must be made to prevent it. Biofilms are formed from planktonic bacteria settling on surfaces. If we control planktonic bacteria, sessile bacteria (biofilms) will therefore be highly controlled [18]. From the time a biofilm forms until it becomes aggressive (i.e., corrosive), there is a time lag. Whatever is to be done must be done within this time lag to prevent aggressive biofilms from forming. This lag time for an organic-rich seawater is more than a week, and in filtered seawater is more than a month [2].

■ Biocide entrance into biofilms obeys diffusion laws. Based on this, one can calculate for a given biofilm thickness how long it will take for the biocide to penetrate into the film [19]. A biofilm with a thickness of 1 mm may allow penetration in about 10 minutes, but the same biofilm that has thickened to 10 mm will allow the same biocide to penetrate in about 20 hours [20].

■ Based on the selection of layup methods, for instance wet layup, it is important to consider the future of the HYD system. If water is to remain within the system, it is highly likely that the water pockets thus formed will constitute spots at which electrochemical/MIC corrosion will start. Keeping in mind that it is impossible to sterilize an industrial system (meaning that no microorganisms, including bacteria, remain) but at best its disinfection is possible (most of the microorganisms are removed but there are still some bacteria inside the system), it follows that these remaining water pockets will induce corrosion either in the form of MIC or other electrochemical processes.

3.2 Closing Remarks

For Group 1 pre-HYD and Group 2 post-HYD risks it is necessary to carefully examine the following issues:

■ If the HYD medium is water from natural sources (such as seawater), care must be taken to draw this water in such a way that both its TDS and nutrients are low. This water, if necessary, must also be filtered to remove the debris normally accompanying such water.

■ HYD water is required to have a uniform temperature distribution. It is advisable that it is allowed to stagnate for a given time before wet parking. During this time, dissolved oxygen will diffuse out, and for wet parking the addition of oxygen scavengers may not be necessary. However, it is advisable to add biocides and inhibitors, as in practice drain/dry is never achieved with 100% efficiency and there is always a risk of leaving water pockets behind in the equipment (a pipe, for example).

■ The prevailing corrosion mechanism in HYD is MIC.

■ Water pockets will act as the necessary electrolyte for electrochemical corrosion and a medium for CRB. Accordingly, although it may look like a "luxury" and be costly, it is good to carry out absence/presence tests for at least five groups of CRB (SRB, SOB, IRB, IOB, and Clostridia). Bear in mind that these tests will only show planktonic bacteria in the HYD water, but this can help a lot in selecting the best possible, broad-spectrum biocide. Also, because at this stage the absence/presence tests will be done by culture-dependent methods, and such methods can reveal only a small percentage of existing bacteria in the environment, one has to be mindful of these intrinsic limitations.

■ If HYD is to be carried out in an welded pipe, care must be given to the quality of the weld in addition to HYD.

■ The material of the pipe is also important. If instead of carbon steel one uses stainless steel, then one must be very careful about chloride concentration and pipe temperature so as not to cause the initiation of pitting and crevice corrosion on the material.

■ HYD is an important test that measures both strength and leakage (contrary to the pneumatic test, which measures leaks only). The time required to carry out HYD must be allowed, and for no reason must this be shortened.

■ The layup, if the time interval between initial water flooding for HYD and putting into service exceeds 30 days, must be considered. Wet layup must be done with clean water that has been treated with a suitable cocktail of biocide + inhibitor + oxygen scavenger. In this regard, the synergy between the chemicals must be considered.

Annex 1: Quantification of MIC Risk Factors Based on Post- and Pre-HYD

In a quantified world, as it is today, perhaps the best way to describe a phenomenon is by mathematical expressions. Here we will present a very simple mathematical way to quantify the risk of MIC that can result from HYD. The "algorithm" consists of the following steps:

- *Step 1:* A list of possible parameters contributing to MIC based on Group 1 and Group 2 risk factors (see Chapter 1, Section 1.3) with arbitrary weights.
- *Step 2:* Classification of the risk of MIC.

Like all other mathematical expressions of a natural phenomenon, there are assumptions to make the math easier. In our approach, we assume the following: (1) the material is carbon steel with no lining and (2) the bacteria are all mesophilic, that is, they can grow and be active within normal ambient temperatures.

Step 1: Based on the pre- and pro-HYD risks and the two mentioned assumptions, we have identified and defined six factors whose importance in terms of the assigned weights will determine if they are important for pre-HYD activities or post-HYD activities. In fact, an MIC factor (shown as F_{MIC}) is defined based on two sets of functions: $F_{HYD\,MIC}$, which defines the factors that could lead into and support MIC during HYD, and $F_{Ser\,MIC}$, which is important to lead into or support MIC during service and thus post-HYD activities. Table A1.1 shows the related items.

TABLE A1.1 Weights Assigned to the Parameters That Can Define Microbial Corrosion Factors According to Group 1 and Group 2 Risk Items*

| Parameter | | F_{MIC} | | Note |
		$F_{HYD\ MIC}$	$F_{Ser\ MIC}$	
HYD water	Distilled demineralized water	0	–	Because of low nutrients, either there are no bacteria or the levels are too low.
	Drinking water	0–3	–	The default value is 3. Experimentation will determine if another value is valid.
	Lake/river/ brackish water/seawater	1–3	–	1 for river, 2 for lake, 3 for brackish water, 2 for seawater.
TDS	Filtered less than 50 microns	0–3	–	If we can filter below 1 micron, then the value will be 0. Otherwise, based on the filter mesh, 1–3 will be the value. This is because, based on filter size, nutrient levels and sediments will be lower.
	Unfiltered or filtered greater than 50 microns	3	–	Very high likelihood for nutrient ingress for bacteria and sedimentation.
Temperature (°C)	$T < 0, T > 82$	0	0	Nil likelihood of active mesophilic CRB under normal working conditions.
	0–15.5	0	0	While active, it is not that likely that CRB activity will be significant.

(Continued)

TABLE A1.1 (*Continued*)   **Weights Assigned to the Parameters That Can Define Microbial Corrosion Factors According to Group 1 and Group 2 Risk Items***

Parameter		F_{MIC}		Note
		$F_{HYD\ MIC}$	$F_{Ser\ MIC}$	
	46–82	2	2	While too hot for the CRB types of concern, the temperature may speed up thermophilic growth.
	15.5–46	3	3	The best temperature range for most CRB.
Working conditions	Dry gas	–	1	The so-called "dry" gas that has a certain level of water induced during long utilization periods, or high volumes of the gas that already has some moisture content.
	Wet gas	–	2	High likelihood to create a suitable electrolyte as well as creating a good venue for biofilming.
	Long wet parking, water-flooded pipes, or appropriate biocide has not been applied	3	3	High likelihood to create a suitable electrolyte as well as creating a good venue for biofilming.

(*Continued*)

TABLE A1.1 (Continued) **Weights Assigned to the Parameters That Can Define Microbial Corrosion Factors According to Group 1 and Group 2 Risk Items***

Parameter		F_{MIC}		Note
		$F_{HYD\ MIC}$	$F_{Ser\ MIC}$	
Sessile bacteria	SRB > 10^2 counts/cm^2	3	3	The aim is to always keep the number of bacteria less than 10^2 counts/cm^2, otherwise cession risk exists. However, for the particular case of SRB, as there is no known link between their numbers and the corrosion rate, even if their numbers are less than 10^2 counts/cm^2, risk of MIC exists.
	SRB < 10^2 counts/cm^2	1	1	
	GAB > 10^2 counts/cm^2	3	3	
	GAB < 10^2 counts/cm^2	1	1	
	GAnB > 10^2 counts/cm^2	3	3	
	GAnB < 10^2 counts/cm^2	1	1	It must be noted that even if the number of bacteria is less than 10^2 counts/cm^2, there is still a small likelihood for initiation or enhancement of corrosion based on the internal chemistry of the system and its hydrodynamics (e.g., water stagnancy).

(Continued)

TABLE A1.1 (Continued) Weights Assigned to the Parameters That Can Define Microbial Corrosion Factors According to Group 1 and Group 2 Risk Items*

Parameter		F_{MIC}		Note
		$F_{HYD\ MIC}$	$F_{Ser\ MIC}$	
Planktonic bacteria	SRB > 1^2 counts/cm^2	3	3	The goal must be to keep the number of these bacteria less than 10^4 counts/cm^2. However, for the particular case of SRB and lack of any relationship between their numbers and corrosion rate, the ideal goal is to set the number of SRB to zero. If, from a practical point of view, this is not achievable, the number of planktonic SRB must be as low as possible as it is always possible for these planktonic bacteria to turn into the sessile state and create risk. Therefore, the "less than 10^4 counts/cm^2" is not ideal, but practically speaking an MIC risk-lowering factor.
	SRB < 11 counts/cm^2	0	0	
	GAB > 10^4 counts/cm^2	3	3	
	GAB < 10^4 counts/cm^2	1	1	
	GAnB < 10^4 counts/cm^2	1	1	

(Continued)

TABLE A1.1 (Continued) **Weights Assigned to the Parameters That Can Define Microbial Corrosion Factors According to Group 1 and Group 2 Risk Items***

GAB, general aerobic bacteria; GAnB, general anaerobic bacteria.
*Darwin, Annadorai, and Heidersbach [2] have used a seemingly similar algorithm. We found that approach is useful but not completed. In our approach, we have added more factors and have also applied different weights to the assigned parameters. While we appreciate these authors and their pioneering work, we would like to emphasize upon the fundamental differences that exist between our approaches.

TABLE A1.2   Classification of the Risk of MIC Due to Risk Groups Associated with HYD

Value for the Probability of the Risk of MIC	Description of the MIC Risk
≤3	Negligible
≤6	Low
≤12	Moderate
≤24	High
>24	Very high

Step 2: The probability of the risk of MIC due to both $F_{\text{HYD MIC}}$ and $F_{\text{Ser MIC}}$ and wet parking is defined as outlined in Table A1.2.

EXAMPLE 1

The HYD water is unfiltered pond water. The temperature is 20°C and the pipe will pass dry gas. The wet parking time is one week. Based on the values given in Table A1.2, the risk of MIC is "High."

EXAMPLE 2

The HYD water is seawater and filtered less than 50 microns but over 1 micron. The temperature, wet parking time, and passing fluid conditions are the same as in Example 1. This time the risk of MIC is "Moderate."

Annex 2: A List of Biocides and Their Pros and Cons

The chemicals presented here as biocides are capable of killing bacteria (hence the name "bio-cide"). However, all of them have pros and cons [7]. In addition to their intrinsic limitations, one has to be aware of the possible antagonistic effects created by adding these chemicals to other chemicals in the equipment. An example of such antagonism was given in section 3.1.4.1. and we will not repeat it here. It must also be noticed that due to the fact that biocides cannot differ between target organisms with other organisms that may not harmful from a corrosion point of view, use of biocides can have serious environmental challenges.

Chemical management is a serious issue when dealing with the management of post-HYD corrosion risks. In the absence of a proper chemical management plan, if anodic inhibitors (such as nitrates) are added to an already microbially contaminated system, then the inhibitor will in fact be used as food for the bacteria. It is also equally important to note that biofims, while still thin and in the early stages of formation, are much easier to tackle. When they mature, it may not be sensible to treat the MIC problem with only a chemical option; a combination of this and physical–mechanical treatment (such as pigging) must also be used with a biocide. Table A2.1 below can assist in determining what biocide could be useful for a given system based on the biocide's advantages and disadvantages [7].

TABLE A2.1   A Brief on Advantages and Disadvantages of Some Frequently Used Biocides

Chlorine

(+):

- Economical
- Broad-spectrum activity
- Effective
- Monitoring dosages and residuals is simple

(–):

- Hazard concerns for the operator
- Ineffective against biofilm bacteria
- Ineffective at high pH
- Inactivation by sunlight and aeration
- Corrosive to some metals
- Adverse effect on wood
- Feeding (dosing) equipment is costly and requires extensive maintenance
- Limitations imposed by environmental authorities on the discharge of chloramines and halomethanes

Chlorinating Compounds (Bleach [NaOCl], Dry Chlorine [Ca (OCL)$_2$])

(+):

- Circumvent the danger of handling chlorine
- As effective as chlorine

(–):

- Can cause scaling problems
- Expensive
- Larger quantities needed than when using gaseous chlorine

Chlorine Dioxide (ClO$_2$)

(+):

- pH insensitive
- Good oxidizing agent for biomass
- Tolerates high levels of organics
- Dissolves iron sulfides

(–):

- Special equipment is required for generation and dosing
- Toxic
- Expensive

(Continued)

TABLE A2.1 *(Continued)* **A Brief on Advantages and Disadvantages of Some Frequently Used Biocides**

Chloramines (Such as Ammonium Chloride)

(+):

- Good biofilm activity
- Good persistence in long distribution systems
- Has reduced corrosivity
- Low toxicity

(−):

- Ammonia injection is required
- Costs more than chlorine alone
- Poor biocidal properties compared to free chlorine

Bromine

(+):

- More effective than chlorine at a higher pH
- Broad-spectrum activity on bacteria and algae over a wider pH range than hypochlorous acid
- Bromamines are environmentally less objectionable and less reactive with hydrocarbons, and so on, reducing the production of halomethane

(−):

- Similar to chlorine compounds
- Expensive

Ozone

(+):

- A natural biocide, effective as a detachment agent against sessile bacteria on stainless steel surfaces
- Resembles advantages of chlorine
- Non-polluting and harmless to aquatic organisms

(−):

- Like chlorine, it is affected by pH, temperature organics, and so on
- Its oxidizing effect does not persist throughout the system, so ozone is used in small systems or specific sites within larger systems
- Ozone must be generated on-site, requiring investment for installation and running the equipment

(Continued)

TABLE A2.1 *(Continued)* **A Brief on Advantages and Disadvantages of Some Frequently Used Biocides**

Sodium and Hydrogen Peroxides

(+):
• Used as a sanitizing agent
• Have many of the same advantages as ozone

(–):
• Requires high concentrations and extensive contact time (to kill the microorganisms)
• Cheaper and more safe than ozone
• Careful use not to stimulate corrosion

Aldehydes

1. *Formaldehyde (HCHO)*

(+):
• Economical

(–):
• Suspected of being a carcinogen
• High dosages are required
• Reacts with ammonia, hydrogen sulfide, and oxygen scavengers

2. *Glutaraldehyde*

(+):
• Broad-spectrum activity
• Relatively insensitive to sulfide
• Compatible with other chemicals
• Tolerates soluble salts and water hardness

(–):
• It is deactivated by ammonia, primary amines, and oxygen scavengers

3. *Acrolein*

(+):
• Broad-spectrum activity
• Penetrates deposits and dissolves sulfide constituents
• In highly contaminated waters, it is generally more economical/cost-effective than chlorine
• No particular environmental hazards

(–):
• Difficult to handle
• Reactive with polymers and scavengers, and violently reacts with strong acid and alkalis
• Potentially flammable
• Highly toxic to humans

(Continued)

TABLE A2.1 (*Continued*)   A Brief on Advantages and Disadvantages of Some Frequently Used Biocides

Amine-Type Compounds

1. *Quaternary amine compounds*

(+):
- Broad-spectrum activity
- Good surfactancy
- Persistence
- Low reactivity with other chemicals

(–):
- Inactivated in brines
- Foaming
- Slow acting

2. *Amine and diamine*

(+):
- Broad-spectrum activity
- Have some inhibition properties
- Effective in sulfide-bearing waters

(–):
- React with other chemicals, particularly anionics
- Less effective in waters with high levels of suspended solids

Halogenated Compounds

1. *Bronopol*

(+):
- Broad-spectrum activity
- Low human toxicity
- Ability to degrade

(–):
- Available as a dry chemical
- Breaks down in high pH

2. *DBNPA*

(+):
- Broad-spectrum activity
- Fast acting and effective (at pH above 8 it must be used for quick kill situations)
- No apparent difficulties related to effluent discharge when these materials are applied as recommended

(*Continued*)

TABLE A2.1 (*Continued*)   **A Brief on Advantages and Disadvantages of Some Frequently Used Biocides**

(–):

- Expensive
- Affected by sulfides
- Must be adequately dispersed to ensure effectiveness due to low solubility in water
- Although effective against bacteria at low concentrations, higher concentrations are required to control most algae and fungi, making them less cost-effective

Sulfur Compounds

1. *Isothiazolone*

(+):

- Broad-spectrum activity
- Compatible with brines
- Good control of many aerobic and anaerobic bacteria (like anti-sessile bacteria) and have activity against many fungi and algae at acidic to slightly alkaline pH values
- Low dosages are required
- Degradable

(–):

- Cannot be used in sour systems
- Expensive
- Less cost-effective when the system contains significant amounts of sessile or adhering biomass; in such cases, the use of a penetrant/ biodispersant enhances the effectiveness of the biocide
- Extreme care required because of potential adverse dermal effects; automated feeding systems are strongly recommended

2. *Carbamates (alkyl thiocarbamates)*

(+):

- Effective against SRB and spore formers
- Effective in alkaline pH
- Useful for polymer solutions

(–):

- High concentrations are required

3. *Metronidazole (2-methyl-5 nitroimidazole-1-ethanol)*

(+):

- Effective against SRB
- Compatible with other chemicals

(*Continued*)

TABLE A2.1 *(Continued)*   A Brief on Advantages and Disadvantages of Some Frequently Used Biocides

(–):

- It is specific to anaerobic organisms

Quaternary Phosphonium Salts (QUATS)

(+):

- Broad spectrum of killing activity and good stability; they are generally most effective against algae and bacteria at neutral to alkaline pH
- Low toxicity
- Stable and unaffected by sulfides

(–):

- Not effective fungicides at any pH
- Their activity is mostly reduced by high chloride concentrations, high concentrations of oil and other organic foulants, and by accumulations of sludge in the system
- Excessive overfeed of some types of QUATs may contribute to foaming problems, especially in open recirculating systems with organic contamination

Annex 3: Corrosion Knowledge Management*

Corrosion professionals are collectively interested in technical matters related to corrosion management, issues such as the best practice for materials selection, cathodic protection (CP), coatings, and so on. However, there is another important issue in the management of corrosion: the role played by the human factor. Our industrial experience suggests there are at least two reasons why the importance of the human factor and management is not considered in corrosion practice:

- Corrosion is always considered by top and middle managers as a too technical matter, best left to the "professionals."
- Corrosion professionals themselves also see their corrosion practice as too technical to be communicated with a language other than "corrosion language."

These two issues have led to a "China wall" between corrosion professionals and their non-corrosion managers. We intend to find a common language for managers and corrosion professionals—*corrosion knowledge management*—or CKM (Figure A3.1).

A3.1 What Is Corrosion Management?

Before we explain corrosion management (CM), we have to define what is meant by engineering importance.

---

* This is a modified form of the paper by R. Javaherdashti, Corrosion knowledge management for managers, *Materials Performance (MP)*, 55(9), 2016.

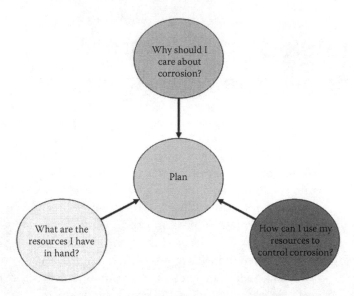

FIGURE A3.1 Three important questions every manager must answer. (Adapted from R. Javaherdashti, Corrosion knowledge management for managers, *Materials Performance* (*MP*), 55(9), 2–5, 2016.)

"Engineering importance" can be defined as a function of both "cost" and "risk." Risk, itself, is a function of "likelihood" and "consequences." Therefore, engineering importance can be defined as

$$\text{Engineering Importance (EI)} = \text{Cost} \times \text{Risk}$$

Corrosion risks can be determined using methods related to integrity management and its practice. The outcome of such studies (based on risk-based inspection [RBI] and then feeding the findings into an integrity model to come up with an action plan, and then determining what to do with regard to corrosion) are all parts of what one may call "corrosion management." In this approach, one may identify the "threats" and, based on that, figure out an action plan. In all CM plans, there is no detailed description of the management factor.

Therefore, in rough terms, one can identify CM as "Any technical approach taken by corrosion professionals to control corrosion by lowering its risk." The main aim of any CM is not defining the economic or ecological cost of corrosion. CM uses these data to justify the consequences that could be expected should the likelihood of corrosion not be observed.

A3.2 What Is Corrosion Knowledge Management?

What is corrosion knowledge management? In simple terms CKM can be defined as "a function-based, result-driven information exchange system between senior managers and industry experts." CKM seeks to find answers to the following three questions:

- Does the manager know his resources?
- Does the manager have a clear idea about his target?
- Does the manager know how to set the shortest and most feasible way to reach his targets from his at-hand resources?

In order for a manager to make decisions about management of corrosion in his plant, he has to answer the "why, what, and how" questions shown in Figure A3.1.

When the manager is prepared to answer these questions, then naturally he will wonder what his resources are and how he can reach his goal for controlling corrosion. Figure A3.2 briefly shows the resources a manager can have at hand.

There are seven resources that any manager can have available. The manager must first identify why corrosion is important. It is at this stage (defining why corrosion is important) that any economic/ecological data will be essential. Based on this, then, the manager must redefine his resources. If based on the facts and figures, MIC is of relative importance in his system, then he should start to plan for all seven items he has as his resources. If he has no MIC expert, he must employ one. If he cannot afford to do so, he must plan training

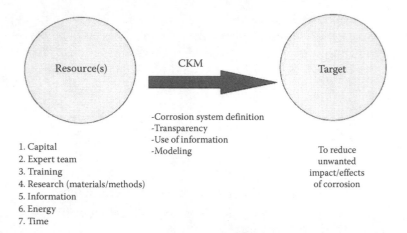

1. Capital
2. Expert team
3. Training
4. Research (materials/methods)
5. Information
6. Energy
7. Time

FIGURE A3.2 Managerial resources that must be redefined based on a CKM approach. (Adapted from R. Javaherdashti, Corrosion knowledge management for managers, *Materials Performance* (*MP*), 55(9), 2–5, 2016.)

to educate and re-educate his personnel in relation to MIC.

Based on this need, he may want to spend money on MIC research, and all pieces of the puzzle will fall into place and he will complete the picture.

A3.3 A Brief Description of CKM Principles

Unlike CM approaches where the corrosion professional has the first and last word, according to a CKM approach, a corrosion professional has to work closely with a manager to define not only the technicalities (such as defining the corrosion system) but also the three CKM principles that deal with corrosion from an organizational–managerial point of view.

CKM is a method that allows the use of in-hand resources in a more profitable way to mitigate corrosion by lowering its costs. It has four components:

■ Modeling
■ Use of information

■ Transparency
■ Corrosion system definition (COFS/CINS)

We will briefly explain just two components: "corrosion system definition" and "use of information."

A3.3.1 Components of CM as a Managerial Tool: Corrosion System Definition

It is worth concentrating more on corrosion system definition. An important aspect of solving corrosion problems is understanding the system in which corrosion is taking place. A corrosion system is defined as "a part of the universe in which corrosion occurs and is of interest to us." The corrosion system can be considered to consist of subsystems such as A, B, C, and so on. If the corrosion problem of each subsystem i is shown as corr(i) and corrosion types observed in each subsystem as a, b, c, ..., one can write

$$\text{corr(A)} = \{a_1, a_2, a_3, ..., a_n\}$$
$$\text{corr(B)} = \{b_1, b_2, b_3, ..., b_n\}$$
$$\text{corr(C)} = \{c_1, c_2, c_3, ..., c_n\}$$

Corrosion of a system (COFS) is then defined as

$$\text{COFS} = \text{corr(A)} \cup \text{corr(B)} \cup \text{corr(C)}$$

Corrosion in a system (CINS) is then defined as

$$\text{CINS} = \text{corr}(i) \cap \text{COFS}$$

As an example, take corrosion in an automobile. Suppose we define a car as a corrosion system that has typical types of corrosion in its subsystems. In this case, the subsystems can be defined as A = chassis, B = fuel system, C = brake system.

Corrosion problems in each of the above subsystems, with important mechanisms in parentheses, can be shown as

corr(A) = {uniform, pitting, (crevice), fretting, stress cracking, ...}
corr(B) = {(pitting), crevice, coating failure}
corr(C) = {(pitting), crevice, galvanic, (fretting), (coating failure)}

So, a project with the goal of solving ALL corrosion problems in a car would have to deal with ALL of the corrosion problems in ALL subsystems; in other words, it would be a COFS approach. In this case, study of the corrosion of just a given subsystem, such as corr(C), will be a CINS approach.

It is important to distinguish between CINS and COFS approaches, because, otherwise, many problems such as expected time span of the project or required capital for doing the project may arise. An important point is that, in practice, it is better for a corrosion expert to define a corrosion system as the system with "highest risk." More often than not, a large percentage of the risk (>80%) is found to be associated with a small percentage of the equipment (<20%). Once identified, the higher-risk equipment becomes the focus of inspection and maintenance to reduce the risk, while opportunities may be found to reduce inspection and maintenance of the lower-risk equipment without significantly increasing risk. In other words, to be on the safe side, it is better to choose as the system of concern the one with higher risk and define COFS and CINS according to the real, working conditions of the system.

A3.3.2 Components of CM as a Managerial Tool: Use of Information

"Information" has a broad meaning. However, in the context of CKM, information will be referred to in terms of the two following concepts:

■ White information
■ Dark information

Information that is accumulated and processed within an organization about corrosion and is applied is called "white information." However, if this information is just processed and stored and is not applied anywhere, we call it "dark information." In any organization dealing with corrosion, there is a good deal of white information, such as, but not limited to, standards, procedures, recommended practices, and so on. However, there is an even greater amount of information, even in the form of written information, that is not used—this is "dark information." Examples of dark information can be the field experiences of experienced corrosion professionals of the organization, histories of communication regarding a case of corrosion, and so on. Most corrosion problems in practice happen because the top management is inclined to ignore dark information. A review of corrosion-related disasters, such as the events around the All-American operated pipeline (2015), the oil spill in Kalamazoo River (Michigan, July 2010), and the explosion of NDK Crystal (2009), as well as similar corrosion cases around the world, were all found to be a consequence of ignoring dark information that was already present in the organization/company.

A3.4 Application of CKM Principles to Environmental Issues

We need to borrow two terms from ISO 14000 terminology: "environmental aspect" and "environmental impact." In its most simple terms, "environmental aspect" relates to all activities, functions, products, or services of an organization (factory, plant) that are capable of interaction with the environment, whereas "environmental impact" is any change in environment, useful or harmful, that may have resulted, partly or generally, from activities, products, or services of an organization. For example, dust resulting from industrial activities such as sandblasting is an example of an environmental aspect, whereas the problems it causes for health are the environmental impact.

It is to be noted that, according to ISO 14000, one should always consider the most dangerous impact as it is not practical to investigate and control all environmental impacts. In this way, ISO 14000 resembles CM, where "a corrosion system, by a corrosion expert, is defined as the system with *highest risk*." So, when considering environmental impacts, one should not only take care of the most dangerous impact from an environmental point of view, but one must also consider the one that is of highest risk of corrosion.

The proposed method to study environmental problems from a corrosion viewpoint is to define the environmental aspect of the case under study as corrosion. Its environmental impacts can then be divided into two subgroups:

- Direct environmental impacts of corrosion
- Indirect environmental impacts of corrosion

Direct environmental impacts of corrosion are those impacts that have resulted just because of corrosion, leakage of dangerous liquids or gases due to weld decay, or stress corrosion cracking (SCC), for example. Indirect environmental impacts of corrosion are those impacts where, at first glance, corrosion does not seem to be the main cause. One example of this could be energy waste in a car because of a malfunction, which itself could have been the result of corrosion.

The study of direct environmental impacts of corrosion could be rather simple, because one would have to define just the three following items clearly:

- Definition of corrosion system (CINS or COFS)
- Recognizing the environmental aspect, that is to say, the corrosion type
- Defining and recognizing the environmental impact and determining its domain

To study the above three items, the environmental expert has to work with a corrosion expert to be able to define the system (corrosion system) and the corrosion type. It is even possible that the corrosion expert will be able to define the domain in which the environmental impact of that type of corrosion will take place. One example is the study of atmospheric corrosion in industrial areas and comparing it with atmospheric corrosion in tropical areas.

The study of indirect environmental impacts of corrosion could also focus on three items:

- Definition of corrosion system (CINS or COFS)
- Recognizing the environmental aspect, that is to say, the corrosion type
- Defining and recognizing the environmental impact and determining its domain

The main difference between studying direct and indirect environmental impacts of corrosion, however, is that to investigate indirect environmental impacts of corrosion, the following items must also be considered:

- Assessment of energy loss in the system by applying energy loss assessment methods (note that the definition of "system" will be crucial in this regard)
- Determining the real depreciation rate of the system by looking at the system function (the value of depreciation may not be the same as that resulting from financial studies due to the very nature of the system)
- Investigation of corrosion type
- Studying the maintenance of the system
- Monitoring and measuring corrosion in the system

Another difference is that the domain of indirect environmental impacts of corrosion can be broader than that of direct environmental impacts. The main reason is that the factors contributing to the environmental impacts of corrosion are not limited to corrosion alone, and many other factors may also be contributing. For instance, when considering the environmental impacts of corrosion due to failure of a pipeline, the quality of welding and post-welding processes, the soil around the pipe (for a buried pipe), the quality of coatings and linings, the topography of the place where the pipe is located, even the type of fertilizers used nearby (to see if a certain type of corrosion called microbiological corrosion can be stimulated), and things of this sort may also be important.

It can be seen that the domain of environmental impacts of corrosion is so broad that the environmental analyst would not only need the help of a corrosion expert but also an energy auditor to help build the picture in its broadest way.

A3.5 Conclusions

■ Preserving and managing the environment in terms of energy and material resources is important to developed countries. Although methods for energy management have been developed, no method is available yet to manage material.

■ Corrosion knowledge management (CKM) can be regarded as a method to mitigate corrosion and therefore minimize material loss.

■ The principles of CKM can not only be employed by corrosion experts but also by non-corrosionists, and especially managers with little or no background in corrosion, to find the most suitable way to manage resources.

■ Considering corrosion type as an environmental "aspect" and its results for the environment as "impacts," one can distinguish between two types of environmental impacts of corrosion: direct and indirect. Both types of impact have the same elements (defining the corrosion system, defining the environmental aspect or corrosion type, and defining and determining the impact of corrosion on the environment and its domain).

■ Due to the multidimensional nature of the impacts of corrosion on the environment, a multidisciplinary team containing experts in corrosion and energy auditing must help the environmental expert to define the environmental aspects and impacts of a corrosion system, clearly and thoroughly.

■ CKM has the capability of defining corrosion problems within an ecological context. An example of such cases can be HYD and its environmental impact as the HYD water is discharged into the soil or seawater.

References

1. R. Javaherdashti, F. Akvan, On the link between future studies and necessity of including corrosion in a desired future scenario: Presenting a model, *International Journal of Engineering Technologies and Management Research*, 2(4), 1–8, 2015.

2. A. Darwin, K. Annadorai, K. Heidersbach, Prevention of corrosion in carbon steel pipelines containing hydrotest water—An overview, Paper No. 10401, NACE 2010, USA, 2010.

3. J.P. Sinha, C.P. Varghese, *Assessment of seam integrity of an aging petroleum pipeline constructed with low frequency ERW line pipes*, 6th Pipeline Technology Conference 2011, Germany, 2011, http://www.pipeline-conference.com/abstracts/assessment-seam-integrity-aging-petroleum-pipeline-constructed-low-frequency-erw-line.

4. R. Javaherdashti, MIC myths: Avoiding common pitfalls in the practice of hydrotesting and likelihood of microbial induced corrosion, *Corrosion Management*, January–February, 11–14, 2009.

5. R. Javaherdashti, Enhancing effects of hydrotesting on microbiologically influenced corrosion, *Materials Performance (MP)*, 42(5), 40–43, 2003.

6. R. Javaherdashti, C. Nwaoha, H. Tan, *Corrosion and Materials in the Oil and Gas Industries*, CRC Press/Taylor & Francis, USA, 2013.

7. R. Javaherdashti, *Microbiologically Influenced Corrosion—An Engineering Insight*, 2nd edition, Springer, UK, 2017.

8. F.M. Al-Abbas, A.E. Kakpovbia, D.L. Olson, B. Mishra, J.R. Spear, *Biodiversity associated with sweet crude and seawater injection systems: Implication for microbiologically influenced corrosion*, 15th Middle East Corrosion Conference & Exhibition, Paper No. 14106, BSE-NACE, Manama, Kingdom of Bahrain, February 2–5, 2014.

9. R. Javaherdashti, Corrosion and biofilm, in H. Kanematsu and D.M. Barry (eds.), *Biofilm and Materials Science* Springer, Switzerland, 2015.

10. L.J. Korb, D.L. Olson et al., *ASM Handbook, Vol. 13, Corrosion*, ASM International, USA, 1987.

11. J.T. Kochelek, *The Chemistry of Oxygen Corrosion in Wet Pipe Fire Sprinkler Systems and Wet Pipe Nitrogen Inerting (WPNI) for Corrosion Control*, Engineered Corrosion Solutions, LLC, 2015.

12. *U.S. Geological Survey Dissolved Oxygen Tables*, http://water.usgs.gov/software/DOTABLES/

13. J.E. Penkala, J. Fichter, S. Ramachandran, Protection against microbiologically influenced corrosion by effective treatment and monitoring during hydrotest shut-in, Paper No. 10404, NACE 2010, USA, 2010.

14. T.M. Willams, The environmental fate of oil and gas biocides: A review, Paper No. 3876, Corrosion 2014, USA, 2014.

15. P.D. Gilbet, J.M. Grech, R.E. Talbot, M.A. Veale, K.A. Hernandez, Tetrakis hydroxymethyl phosphonium sulfate (THPS) for dissolving iron sulphides downhole and topside—A study of the chemistry influencing dissolution, Paper No. 02030, Corrosion 2002, USA, 2002.

16. O. Olabisi, A.R. Al-Shamari, S. Al-Suleiman, A. Jarragh, A. Mathew, The role of bacterial population density in wet and dry crude asset integrity, Paper No. 5534, Corrosion 2015, USA, 2015.

17. R. Prasad, Chemical treatment options for hydrotest water to control corrosion and bacterial growth, Paper No. 03572, Corrosion 2003, USA, 2003.

18. R. Javaherdashti, Ten commandments to prevent microbiologically influenced corrosion in your system, *Materials Performance (MP)*, 44(11), 46–47, 2005.

19. R. Javaherdashti, F. Akvan, When is too late to inject a biocide? A short introduction to a long story, *International Journal of Research—GRANTHAALAYAH*, 4(5), 1–6, 2016.

20. P.S. Stewart, Diffusion in biofilms, *Journal of Bacteriology*, 185(5), 1485–1491, 2003.

21. R. Javaherdashti, Corrosion knowledge management for managers, *Materials Performance (MP)*, 55(9), 2–5, 2016.

Abbreviations

AMP	adenosine monophosphate
APB	acid-producing bacteria
ATP	adenosine triphosphate
cATP	cellular ATP
CINS	corrosion in system
CKM	corrosion knowledge management
COFS	corrosion of system
CP	cathodic protection
CRB	corrosion-related bacteria
DGGE	denaturing gradient gel electrophoresis
DSA	deposit and surface analysis
DW	demineralized water
EMIC	electrical microbiologically influenced corrosion
FISH	fluorescence in situ hybridization
GAB	general aerobic bacteria
GAnB	general anaerobic bacteria
HIC	hydrogen-induced cracking
HPSC	high-purity steam condensate
HYD	hydrotest, hydrostatic testing
IOB	iron-oxidizing bacteria
IRB	iron-reducing bacteria
LF-ERW	low-frequency electric resistance welding
MDM	microbial diagnostic microarray
MIC	microbial corrosion or microbiologically influenced corrosion
MMM	molecular microbiology methods
MPN	most probable number
MSDS	material safety data sheet
NGS	next-generation sequencing
NR-SOB	nitrite-reducing and sulfur-oxidizing bacteria
PCR	polymerase chain reaction
PWT	post-welding treatment
QGOM	quench-gone organic modified
qPCR	quantitative PCR

QUATS	quaternary phosphonium salts
RBI	risk-based inspection
SAW	submerged arc welding
SCC	stress corrosion cracking
SOB	sulfur-oxidizing bacteria
SRA	sulfate-reducing archaea
SRB	sulfate-reducing bacteria
THPS	tetrakis hydroxymethyl phosphonium sulfate
TOB	thiosulfate-oxidizing bacteria
TRA	thiosulfate-reducing archaea
TRB	thiosulfate-reducing bacteria

Index

Note: Page numbers followed by "*fn*" indicates footnotes.